Rocks and Soil
in the Rocky Mountains

by Barbara Wood

Contents

Science Vocabulary. 4

Mountains 8

Rocks 12

Weathering Changes
Rocks 18

Soil 24

Conclusion 28

Share and Compare 29

Science Career 30

Index 32

Science Vocabulary

soil
Soil is a layer of loose material that covers part of Earth's surface.

Soil is found in mountain valleys.

humus
Humus is a part of soil. It has bits of decayed plants and animals.

Humus gives this soil a dark color.

property
A **property** is something about an object that you can observe with your senses.

Having layers is one **property** of rocks.

texture
Texture is the way an object feels.

This rock has a smooth **texture.**

weathering
Weathering is the breaking apart or dissolving of rocks.

Weathering can make rough rocks smooth.

My Science Vocabulary

- humus
- property
- soil
- texture
- weathering

Mountains

These are mountains. Mountains are tall.

Some mountains are covered with snow.

There are many mountains on Earth. The Rocky Mountains are in the United States and Canada.

This map shows where the Rocky Mountains are located.

These mountains are part of the Rocky Mountains.

The Rocky Mountains are in the western part of North America.

Rocks

Mountains are made of rock.

This man is climbing the rock face of a mountain.

Rocks have different **properties**. Having layers is a property. Some rocks have layers. Others don't.

You can see layers in rocks in Glenwood Canyon in the Rocky Mountains area.

property

A **property** is something about an object that you can observe with your senses.

There are many kinds of rocks in the Rocky Mountains.

Color and **texture** are properties of rocks. Rocks may have a smooth or a rough texture. Having layers is also a property of some rocks.

Properties	Words That Describe Properties
color	orange, white, black
layers	layered, not layered
texture	rough, smooth

different colors    some layers    smooth texture

texture

Texture is the way an object feels.

Minerals can be found in rocks. Minerals have properties, too.

copper

diamond

Minerals can be soft or hard. Soft or hard is a property.

zinc

silver

Weathering Changes Rocks

Weathering can change rocks. Weathering can break apart or dissolve rocks over time. For example, water moves across rocks and breaks off tiny bits.

Over time, water will make these rocks smooth.

weathering

Weathering is the breaking apart or dissolving of rocks.

Wind blows on rocks and breaks off tiny bits, too. Water and wind change the shapes of the rocks.

Wind and water helped shape these rocks.

Water and wind can also move rocks.

Water in mountain streams pushes rocks downhill.

Strong winds blow tiny bits of rock and sand.

Winds blew sand. Over time, this changed the shapes of these rocks near the Rocky Mountains.

Some mountains have glaciers. Glaciers can move rocks.

The ice of the glacier moves downhill.

Sometimes glaciers melt. The rocks from the ice fall to the ground and change shape.

Gravity pulled rocks down in Grand Teton National Park.

Soil

Much of the **soil** in mountain areas is in valleys. Small bits of rock are part of soil.

soil

Soil is a layer of loose material that covers part of Earth's surface.

Humus is part of soil, too.

Humus gives this soil a dark color.

---

humus

Humus is a part of soil. It has bits of decayed plants and animals.

Color and texture are properties of soil. Soil made of very small bits of rocks feels smooth. Soil made of larger bits of rocks feels rough.

Dry red soil

Dry brown soil

Soil near the Rocky Mountains

Look at this soil. How would you describe it? How would it feel?

Conclusion

The Rocky Mountains are made of rock. Valleys between mountains often have soil. Rocks and soil have properties, such as color and texture. Weathering can change rocks to soil over time.

Think About the Big Ideas

1. What can you observe about rocks in the Rocky Mountains?
2. How do rocks change shape?
3. What can you observe about soil?

Share and Compare

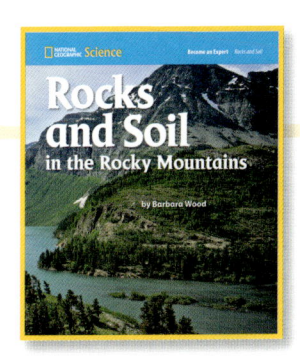

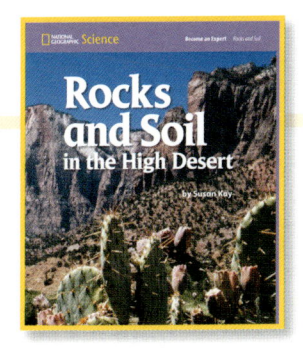

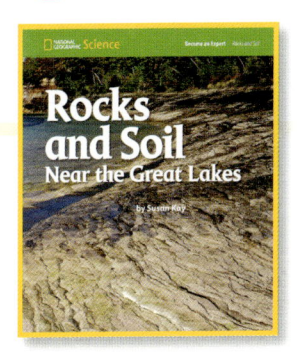

Turn and Talk

Compare rocks and soil in your books. How are they different? How are they alike?

Read

Find your favorite part of the book and read it to a classmate.

Write

Bring a rock to class. Write about its properties. Share your writing with a classmate.

Draw

Draw a picture that shows rocks changing shape. Talk about your drawing with a classmate.

NATIONAL GEOGRAPHIC Science Career

Meet Beverly Goodman

Scientists try to answer questions. They explore nature and make observations. Then they tell people what they find.

Beverly Goodman is a scientist. She studies rocks and soil on coasts, or land by water.

Beverly's team found a thick layer of shells underwater near Israel. Beverly wanted to find out why so many shells were there. By exploring, she and her team found that the layer of shells had pieces of pottery and rounded beach pebbles mixed together. These clues showed that a tsunami must have taken place. A tsunami is a huge wave. It can wash away rocks and soil, and even shells, and carry them to new places.

Index

glacier 22–23

humus 4, 7, 25

property 5, 7, 13, 15–17, 26, 28

soil 4, 7, 24–28

texture 5, 7, 15, 26, 28

weathering 6, 7, 18, 28

Acknowledgments
Grateful acknowledgment is given to the authors, artists, photographers, museums, publishers, and agents for permission to reprint copyrighted material. Every effort has been made to secure the appropriate permission. If any omissions have been made or if corrections are required, please contact the Publisher.

Photographic Credits
Cover (bg) Photodisc/Getty Images; Cvr Flap (t), 5 (t), 13 Richard Wareham Fotografie/Alamy Images; Cvr Flap (c), 19 George Bailey/Shutterstock; Cvr Flap (b), 4 (t), 24 Grant Heilman/Grant Heilman Photography; Title (bg) Buddy Mays/Alamy Images; 2-3, 28 Steve Krull Mountain Scenery Images/Alamy Images; 4 (b), 25 fotoar/Shutterstock; 5 (bl), 15 (l, r) Scientifica/Visuals Unlimited; 5 (br), 16 (l) A. B. Joyce/Photo Researchers, Inc.; 6 (t), 18 Yellow Dog Photography/Alamy Images; 6 (b), 21 John Hoffman/Shutterstock; 7, 14, 22, 23 Marli Miller/Visuals Unlimited; 8-9 Dick Durrance II/National Geographic Image Collection; 11 George Burba/Shutterstock; 12 Gordon Wiltsie/National Geographic Image Collection; 15 (c) Jerome Wyckoff/ Earth Scenes/Animals Animals; 16 (r) Edward Kinsman/Photo Researchers, Inc.; 17 (l) Astrid & Hanns-Frieder Michler/Photo Researchers, Inc., (r) Theodore Gray/Visuals Unlimited; 20 Andre Jenny/Alamy Images; 26 (tl) Russ Bishop/Alamy Images, (tr) Rob Bartee/Alamy Images, (b) Grant Heilman/Grant Heilman Photography; 27 Joel W. Rogers/Corbis; 31 Mark Mallchok; Inside Back Cover (bg) Gerald & Buff Corsi/Visuals Unlimited.

Illustrator Credits
10 Mapping Specialists

Neither the Publisher nor the authors shall be liable for any damage that may be caused or sustained or result from conducting any of the activities in this publication without specifically following instructions, undertaking the activities without proper supervision, or failing to comply with the cautions contained herein.

Program Authors
Kathy Cabe Trundle, Ph.D., Associate Professor of Early Childhood Science Education, The Ohio State University, Columbus, Ohio; Randy Bell, Ph.D., Associate Professor of Science Education, University of Virginia, Charlottesville, Virginia; Malcolm B. Butler, Ph.D., Associate Professor of Science Education, University of South Florida, St. Petersburg, Florida; Nell K. Duke, Ed.D., Co-Director of the Literacy Achievement Research Center and Professor of Teacher Education and Educational Psychology, Michigan State University, East Lansing, Michigan; Judith Sweeney Lederman, Ph.D., Director of Teacher Education and Associate Professor of Science Education, Department of Mathematics and Science Education, Illinois Institute of Technology, Chicago, Illinois; David W. Moore, Ph.D., Professor of Education, College of Teacher Education and Leadership, Arizona State University, Tempe, Arizona

The National Geographic Society
John M. Fahey, Jr., President & Chief Executive Officer
Gilbert M. Grosvenor, Chairman of the Board

Copyright © 2011 The Hampton-Brown Company, Inc., a wholly owned subsidiary of the National Geographic Society, publishing under the imprints National Geographic School Publishing and Hampton-Brown.

All rights reserved. No part of this book may be reproduced or transmitted in any form or by any means, electronic or mechanical, including photocopying, recording, or by an information storage and retrieval system, without permission in writing from the Publisher.

National Geographic and the Yellow Border are registered trademarks of the National Geographic Society.

National Geographic School Publishing
Hampton-Brown
www.NGSP.com

Printed in the USA.
RR Donnelley, Manchester, CT

ISBN: 978-0-7362-7574-3

11 12 13 14 15 16 17

10 9 8 7 6 5 4 3